SpringerBriefs in Fire

Series Editor
James A. Milke
University of Maryland, College Park, USA

More information about this series at http://www.springer.com/series/10476

Kenneth W. Dungan

Incorporating Resiliency Concepts into NFPA Codes and Standards

Kenneth W. Dungan
Performance Design Technologies, Inc.
Knoxville, TN, USA

ISSN 2193-6595 ISSN 2193-6609 (electronic)
SpringerBriefs in Fire
ISBN 978-1-4939-6510-6 ISBN 978-1-4939-6511-3 (eBook)
DOI 10.1007/978-1-4939-6511-3

Library of Congress Control Number: 2016951257

© Fire Protection Research Foundation 2016
This work is subject to copyright. All rights are reserved by the Publisher, whether the whole or part of the material is concerned, specifically the rights of translation, reprinting, reuse of illustrations, recitation, broadcasting, reproduction on microfilms or in any other physical way, and transmission or information storage and retrieval, electronic adaptation, computer software, or by similar or dissimilar methodology now known or hereafter developed.
The use of general descriptive names, registered names, trademarks, service marks, etc. in this publication does not imply, even in the absence of a specific statement, that such names are exempt from the relevant protective laws and regulations and therefore free for general use.
The publisher, the authors and the editors are safe to assume that the advice and information in this book are believed to be true and accurate at the date of publication. Neither the publisher nor the authors or the editors give a warranty, express or implied, with respect to the material contained herein or for any errors or omissions that may have been made.

Printed on acid-free paper

This Springer imprint is published by Springer Nature
The registered company is Springer Science+Business Media LLC New York

Foreword

NFPA's codes and standards address the full spectrum of preparedness, from built-in resiliency in structures and systems to emergency planning and response. In an effort to assess and further define NFPA's position in the realm of resiliency, the Foundation published a report entitled Disaster Resiliency and NFPA Codes and Standards with the purpose to identify those provisions in NFPA codes and standards that embody the concepts of resiliency. Additionally, the project developed an NFPA-centric definition of resiliency and compiled available information to serve as a technical reference for the codes and standards, identifying key gaps in knowledge. Included in the recommendations of the report for future work on resiliency was the development of a book on the incorporation of resiliency concepts into NFPA codes and standards. The guide's scope addresses the contextual definition of resilience, explains the existing resiliency frameworks developed by Federal Agencies, and emphasizes the risk informed approach to applying resiliency concepts to NFPA documents.

Acknowledgments

PROJECT TECHNICAL PANEL
Bill Anderson, The Infrastructure Security Partnership
Morgan Hurley, Aon Fire protection Engineering
Kendal Smith
Nancy McNabb, NIST
Don Bliss, NFPA
Robert Solomon, NFPA
Peter Boynton, Northeastern University
Jim Lathrop, Koffel Associates
Ken Willette, NFPA
Don Schmidt, Preparedness, LLC
Bob Upson, National Fire Sprinkler Association
Allan Coutts, URS Safety Management Solutions LLC
Gary Keith, FM Global

PROJECT SPONSORS
National Fire Protection Association

Contents

1 **Introduction** 1
   1.1 Role of NFPA and Its Documents 1
   1.2 Concepts of Resilience 2
   1.3 Application of Concepts to NFPA Activities 3
   1.4 Purpose and Scope of This Book 4

2 **Overview** 5
   2.1 Background 5
   2.2 Engineered Features and the Built Environment 6
   2.3 Administrative/Operational Features and Emergency Response 7

3 **Risk Informed Approach** 9
   3.1 Risk Metrics 10
   3.2 Risk Identification 11
   3.3 Risk Informed Decision Making 12
       3.3.1 Decision Process 13
       3.3.2 Structural Example 14
       3.3.3 NIST Framework 15
       3.3.4 Decision Trees 16
       3.3.5 Resulting Criteria 18

4 **Buildings and Infrastructure** 19
   4.1 Existing NFPA Codes and Standards 19
       4.1.1 Step 1: Establish Performance Category .. 20
       4.1.2 Step 2: Establish Fire Risk 20
       4.1.3 Step 3 Establish Resilience Criteria 22
   4.2 Future NFPA Documents 23

5 Emergency Preparedness—Planning, Response and Recovery ... 25
- 5.1 Defining Community ... 25
- 5.2 Achieving Community Disaster Resilience ... 26
  - 5.2.1 Planning ... 26
  - 5.2.2 Resources ... 27
  - 5.2.3 Practice ... 30
- 5.3 Recovery ... 32

Appendix 1 Literature Review ... 33
- Definitions of Resilience ... 33
  - Improve Resilience Through Increased Public-Private Partnerships ... 34
- Performance Goals and Objectives ... 38
  - Frameworks for Resilience ... 39

Appendix 2 Recovery Continuum ... 43
- Recovery Continuum – Description of Activities By Phase ... 44

Chapter 1
Introduction

1.1 Role of NFPA and Its Documents

NFPA is a *nonprofit organization devoted to eliminating death, injury, property and economic loss due to fire, electrical and related hazards.* NFPA's Articles of Organization[1] clearly defines this purpose:

> The purposes of the corporation (hereinafter referred to as the Association) shall be to promote the science and improve the methods of fire protection and prevention, electrical safety and other related safety goals; to obtain and circulate information and promote education and research on these subjects and to secure the cooperation of its members and the public in establishing proper safeguards against loss of life and property.

The association pursues this mission by *delivering information and knowledge*. Figure 1.1 illustrates this mission as collecting and disseminating information and developing and sharing knowledge. One significant method available to the association to deliver information and knowledge is the consensus standards process it champions. As a consensus standards organization, NFPA has established processes and procedures to develop health and safety standards. In is broadest sense NFPA defines the term "standard" to include

> a wide variety of technical works that prescribe rules, guidelines, best practices, specifications, test methods, design or installation procedures and the like. The size, scope and subject matter of standards varies widely, ranging from lengthy model building or electrical codes to narrowly scoped test methods or product specifications.[2]

The Technical Committees are given the latitude to develop Codes, Standards, Recommended Practices, or Guides within their scope and under the rules estab-

---

[1] NFPA's Articles of Organization: http://www.nfpa.org/about-nfpa/nfpa-overview/nfpa-operations/articles-of-organization

[2] Definition of standards: http://www.nfpa.org/codes-and-standards/standards-development-process/the-value-of-standards-development-organizations

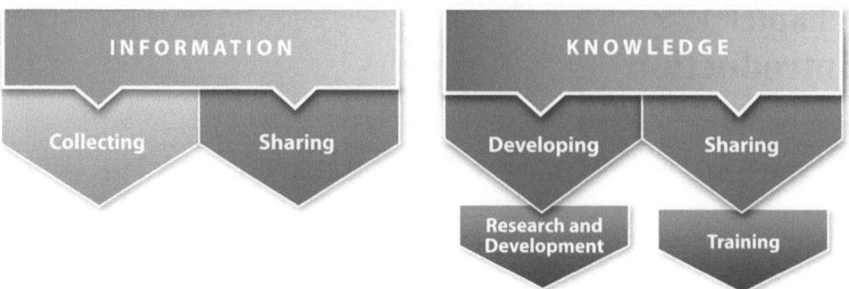

Fig. 1.1 NFPA mission

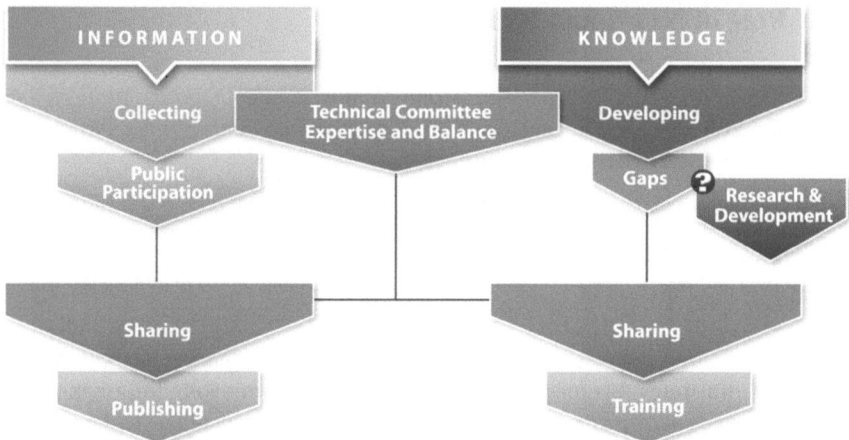

Fig. 1.2 Role of technical committees

lished by NFPA. Figure 1.2 demonstrated how the role of the Technical Committees helps to achieve NFPA's mission.

1.2 Concepts of Resilience

The term *resiliency* has been used with increasing frequency in the context of how we build for, plan for, and respond to the variety of events that could interrupt the desired normalcy. For the purposes of this guide the term is best defined by Presidential Policy Directive (PPD) 21[3]:

---

[3] PPD 21: https://www.whitehouse.gov/the-press-office/2013/02/12/presidential-policy-directive-critical-infrastructure-security-and-resil

1.3 Application of Concepts to NFPA Activities

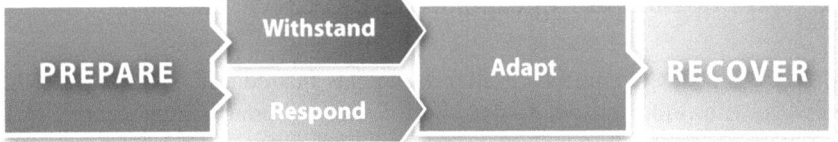

Fig. 1.3 Resiliance

The term "resilience" means the ability to prepare for and adapt to changing conditions and withstand and recover rapidly from disruptions. Resilience includes the ability to withstand and recover from deliberate attacks, accidents, or naturally occurring threats or incidents.

This brief definition includes many key words to focus our attention:

1. *Prepare for Changing Conditions* identifies the need to take certain actions in anticipation of potential events. This anticipation requires some analysis and understanding of the potential event(s).
2. *Adapt to changing conditions* implies a dynamic nature to the events and the responses to those events.
3. W*ithstand disruptions* is one of the goals of preparing for events. The ability or capability to remain unchanged or unimpaired by disruptive events is the most desirable means of achieving resilience.
4. *Recover rapidly from disruptions* is another goal of preparing for and adapting to disruptive events. This *recovery* introduces response activities before, during, and after events to return to the desired state of normalcy in a timely manner.

Figure 1.3 illustrates the relationship between preparedness and recovery.

Appendix 1 to this guide provides a literature review on the subject of *resiliency* to supplement the definition provided in PPD 21[4]. The many definitions and discussions demonstrate the value of considering these concepts in the development of NFPA documents. The common themes from these many definitions can be summarized for their use in characterizing the role of NFPA codes and standards as follows:

1. Resilience includes *technical, organizational, social and economic* dimensions.
2. Resilience requires actions described as planning, preparing, preventing, protecting, mitigating and responding.
3. Resilience requires preparation and response to be adaptive.
4. Resilience should focus on minimizing *damage and disruption to public health and safety, the economy, environment, and national security.*
5. Resilience includes the ability of structures and systems to withstand these external events, whether natural or human-created.

---

[4] https://www.whitehouse.gov/the-press-office/2013/02/12/presidential-policy-directive-critical-infrastructure-security-and-resil

1.3 Application of Concepts to NFPA Activities

NFPA Technical Committees develop documents that educate and inform. When provided with enforcement language, those documents establish societal norms and expectations. The measure of NFPA Technical Committees success is in the quality of its documents. For the purposes of this guide the measures of quality can be stated as Accuracy, Completeness and Timeliness. These are important measure when introducing the evolving concepts of *Resilience*. Committees should consider whether they can achieve these measures of quality when addressing changes to existing documents or the development of new documents.

The decision process for going beyond the "minimum" requirements that traditionally has been the basis for NFPA Codes and Standards, as indicated above, will include *technical, organizational, social, and economic* considerations. These considerations may be difficult to codify. Each Technical Committee will have to evaluate if and how best to serve its documents' users. A starting point is a review of the committee's scope to ensure its activities and deliberations are sanctioned by the Standards Council. Next, the committee should consider the goals and objectives of their current document(s). Are the goals and objectives supported by or in concert with resiliency concepts? Likewise, if NFPA develops documents that contain "how to" advice on the application of resiliency concepts, where is the "when too" advise contained? If the committee desires to develop a new document to support resiliency, the document scope should be approved by the Standards Council.

1.4 Purpose and Scope of This Book

The purpose of this brief is to assist NFPA Technical Committees in understanding the concepts of *Disaster Resilience,* and how those concepts may apply to the scope of their committee documents. The guide's scope addresses the contextual definition of resilience, explains the existing resiliency frameworks developed by Federal Agencies, and emphasizes the risk informed approach to applying resiliency concepts to NFPA documents.

Chapter 2
Overview

2.1 Background

Recent catastrophic events such as hurricane Katrina, Super Storm Sandy, numerous wildfires, and the Fukushima tsunami have focused attention on our ability to design for and react to these types of events. In some cases these events are considered *Beyond Design Basis*. Often the design bases presented in codes, standards, and regulations are a reflection of empirical or statistical information that lead to a predicted worst case. These events may be characterized as the hundred year or the thousand year event. As will be discussed in the next section on Risk Methodologies, the consensus evolves as to how frequent and how severe an event to plan for in our designs and responses. The increasing frequency and escalating severity of the natural phenomena have focused attention on *Community Disaster Resilience*. Since we can have little effect of the causes of these events (at least in the near term), it is prudent to evaluate how best to cope with them. As stated above, the focus of this coping should be minimizing *damage and disruption to public health and safety, the economy, environment, and national security*.

As the literature review in Appendix 1 emphasizes, bouncing back from one of these disasters will depend on many *technical, organizational, social, and economic* decisions. The emphasis of PPD 21 is to engage the broadest participation in the planning and executing improvements in *Community Disaster Resilience*. The voluntary consensus standards process is expected to play a supporting role and at time a leading role in the process. For the purposes of this guide, the roles of NFPA documents are considered to require either **Engineered Features** or **Administrative/Operational Features.** This simplification allows a clearer discussion of the scope, goal, and objectives of individual NFPA documents that may contribution to achieving improved resilience.

USA government initiatives have attempted to provide workable frameworks for contributing to improving *Community Disaster Resilience*. In response to PPD-21, Department of Homeland Security (DHS) embarked on two parallel efforts, *National Infrastructure Protection Program (NIPP) 2013, Partnering for Critical*

Infrastructure Security and Resilience (initiated with NIPP 2006 and 2009) and *National Disaster Recovery Framework: Strengthening Disaster Recovery for the Nation,* developed by FEMA in 2011. To support the Risk Analysis aspects of these efforts, *CPG 201: Threat and Hazard Identification and Risk Assessment* was developed. The most recent effort has been orchestrated by NIST and has been documented as *Community Resilience Planning Guide for Buildings and Infrastructure Systems,* (CRPG), NIST Special Publication 1190.

2.2 Engineered Features and the Built Environment

NIPP 2013 tends to focus on **Engineered Features,** but not to the exclusion of **Administrative/Operational Features.** NIPP 2013 clarifies the definitions from PPD 21 for its applications:

Resilience—*The ability to prepare for and adapt to changing conditions and withstand and recover rapidly from disruptions; includes the ability to withstand and recover from deliberate attacks, accidents, or naturally occurring threats or incidents.*

Recovery—*Those capabilities necessary to assist communities affected by an incident to recover effectively, including, but not limited to, rebuilding infrastructure systems; providing adequate interim and long-term housing for survivors; restoring health, social, and community services; promoting economic development; and restoring natural and cultural resources.*

NIPP 2013 provides a clear and useful set of goals regarding the strengthening of the security and resilience of the national critical infrastructure. However, these goals apply equally well to the broader applications of *Community Disaster Resilience*. Likewise, these goals speak to NFPA's mission.

Assess and analyze threats to, vulnerabilities of, and consequences to critical Infrastructure to inform risk management activities;

Secure critical infrastructure against human, physical, and cyber threats through sustainable efforts to reduce risk, while accounting for the costs and benefits of security investments;

Enhance critical infrastructure resilience by minimizing the adverse consequences of incidents through advance planning and mitigation efforts, and employing effective responses to save lives and ensure the rapid recovery of essential services;

Share actionable and relevant information across the critical infrastructure community to build awareness and enable risk-informed decision making; and

Promote learning and adaptation during and after exercises and incidents.

CPG 201: Threat and Hazard Identification and Risk Assessment, Second Edition is referenced in NIPP 2013. The guide describes a four-step process for developing a risk assessment, which will be discussed in the next section. As will be pointed out, the economic burdens of action and inaction mandate decisions based on risk.

NIST's CRPG supports the objectives of NIPP 2013. This ambitious guide contains valuable insights into the role and potential content of consensus codes and standards to support improved resiliency. Chapter 3 of the NIST CRPG clearly states the importance of buildings to community resilience:

> *The built environment is an essential part of community resilience. Social institutions, including family/kinship, education, health, government, economy, media, other community-based organizations, rely on buildings and infrastructure systems at all times—before, during, and after a hazard event occurs. Building clusters (buildings grouped by similar function) and infrastructure systems must be functional to support restoration of neighborhoods, care for vulnerable populations, and restore the community's economy.*

Chapter 4 outlines the establishment of Goal and Objectives for Community Resilience. It explains the interaction of *withstand* and *recovery* aspects of resilience:

> *Desired performance goals depend on: (1) an acceptable level of damage that occurs for a particular hazard level (performance level) and (2) the corresponding recovery time to restore full functionality. Performance levels address life safety and post-event functionality. Recovery times help prioritize repair and reconstruction efforts. Additionally, performance goals should consider the role of a facility or system.*

NFPA codes and standards that address occupancies or systems can include **Engineered Features** that play a role in achieving community resilience goals. The consensus process may also be helpful in establishing methodologies for developing and evaluation objectives and criteria to meet resiliency goals.

2.3 Administrative/Operational Features and Emergency Response

Decisions have to be made and actions have to be taken in order to respond to disasters. The use of the term disaster may be prejudicial, since the transition from event to disaster may be accelerated by poor decisions and improper action or inaction. *National Disaster Recovery Framework: Strengthening Disaster Recovery for the Nation,* developed by FEMA in 2011, provides excellent information on the planning and executing response to event that threaten our communities. The framework discusses the elements of planning, coordinating, communication and acting across local, state, regional and even national jurisdictions. The main purpose of the framework is described as:

> *The National Disaster Recovery Framework (NDRF) applies to all Presidentially-declared major disasters though not all elements will be activated for every declared incident. Many of its concepts and principles are equally valid for non-declared incidents that have recovery consequences. The core concepts as well as the Recovery Support Function (RSF) organizing structures outlined in the NDRF may be applied to any incident regardless of whether or not it results in a Presidential disaster declaration.*

The NDRF includes a very insightful discussion of recovery, in terms of needs and duration that applies to establishing priorities for response and recovery activities.

The recovery process is best described as a sequence of interdependent and often concurrent activities that progressively advance a community toward a successful recovery. However, decisions made and priorities set early in the recovery process by a community will have a cascading effect on the nature and speed of the recovery progress.

This recovery continuum is displayed graphically in NDRF, *FIGURE 1—RECOVERY CONTINUUM—DESCRIPTION OF ACTIVITIES BY PHASE*. This figure is provided in Appendix 2. The recovery continuum is also useful in establishing priorities for **Engineered Features**.

The **Administrative/Operational Features** or programmatic activities as part of response and recovery are both pre-disaster and post disaster. The planning and preparing aspect are intended to be before the need arises. NDRF lists the key principles of *Pre-Disaster Recovery Planning*, as well as *Recommended Activities*. As will be discussed in Chap. 5 of this guide, these **Administrative/Operational Features** require sufficient Planning, Resources and Training to be successful. NFPA codes and standards that address emergency planning and response currently address **Administrative/Operational Features** and can support the community resilience goals.

Chapter 3
Risk Informed Approach

Resources available to achieve resilient infrastructure and improved community disaster resilience are not inexhaustible. Therefore the approaches recommended by the DHS initiatives all emphasize risk as the basis for decisions. NIPP 2013 identifies understanding risk as its *National Plan's* first core tenet:

> Risk should be identified and managed in a coordinated and comprehensive way across the critical infrastructure community to enable the effective allocation of security and resilience resources.

CPG 201: Threat and Hazard Identification and Risk Assessment, Second Edition defines risk and promotes it as the basis for decision making as follows:

> Every community should understand the risks it faces. By understanding its risks, a community can make smart decisions about how to manage risk, including developing needed capabilities. Risk is the potential for an unwanted outcome resulting from an incident, event, or occurrence, as determined by its likelihood and the associated consequences.

The product of likelihood and consequence will be the definition of risk in this section. This is different from hazard identification, since risk analysis attempts to evaluate both the magnitude and frequency of the events. CPG 201 describes a four-step process for developing a risk assessment. These four steps are as follows:

1. ***Identify the Threats and Hazards of Concern***. *Based on a combination of experience, forecasting, subject matter expertise, and other available resources, identify a list of the threats and hazards of primary concern to the community.*
2. ***Give the Threats and Hazards Context***. *Describe the threats and hazards of concern, showing how they may affect the community.*
3. ***Establish Capability Targets***. *Assess each threat and hazard in context to develop a specific capability target for each core capability identified in the National Preparedness Goal. The capability target defines success for the capability.*
4. ***Apply the Results***. *For each core capability, estimate the resources required to achieve the capability targets through the use of community assets and mutual aid, while also considering preparedness activities, including mitigation opportunities.*

Steps three and four reference the *Core Capabilities* identified in the National Preparedness Goal[(x)] documents. For the purposes of this guide, the first two steps represent the basis for Risk Informed decisions related to NFPA activities. A major component of the National Preparedness System is the *Identifying and Analyzing of Risk*, and that component is intended to be supported by CPG 201.

If the decisions on the most effective deployment of limited resources must be based on risk, understanding those risks is an important consideration in the development of codes and standards. This is not a new concept to the NFPA. In March 2007, the Fire Protection Research Foundation published *Guidance Document for Incorporating Risk Concepts into NFPA Codes and Standards*. Although the contexts of this guidance document revolves around fire as the primary risk, the general concepts and the discussions of example methodologies apply very well to the broader applications of *resilience*.

3.1 Risk Metrics

Chapter 5, Risk Criteria of the *Guidance Document for Incorporating Risk Concepts into NFPA Codes and Standards* should be used as a companion to this guide. The risk metrics listed in its Sect. 5.1 are appropriate for NFPA's codes and standards.

- Life Safety (public and worker)
- Property Protection
- Continuity of Operations
- Environmental Protection
- Preservation of Cultural Heritage
- Preservation of National Security

These broad metrics are equally appropriate for the National Preparedness Missions. These missions appear below as extracted from CPG 201:

1. **Prevention:** *Prevent, avoid, or stop an imminent, threatened, or actual act of terrorism.*
2. **Protection:** *Protect our citizens, residents, visitors, and assets against the greatest threats and hazards in a manner that allows our interests, aspirations, and way of life to thrive.*
3. **Mitigation:** *Reduce the loss of life and property by lessening the impact of future disasters.*
4. **Response:** *Respond quickly to save lives; protect property and the environment; and meet basic human needs in the aftermath of a catastrophic incident.*
5. **Recovery:** *Recover through a focus on the timely restoration, strengthening, and revitalization of infrastructure, housing, and a sustainable economy, as well as the health, social, cultural, historic, and environmental fabric of communities affected by a catastrophic incident.*

3.2 Risk Identification

The overarching theme is the health and wellbeing of the population: not just freedom from injury, but health, safety, economic stability and political stability. In that context, property, operations, the environment, and cultural heritage are valued for their contribution to the health and wellbeing of the population. This is an important concept for evaluating the risk-benefit of changes to NFPA documents to improve community disaster resiliency.

The FEMA *Recovery Continuum* (see Appendix 2) also identifies its priorities for timeliness based on public health and wellbeing. The concept of timeliness has limited application to design, but it is critical in addressing the risk metrics of response and recovery.

3.2 Risk Identification

NATURAL	TECHNOLOGICAL	HUMAN-CAUSED
Avalanche	Airplane Crash	Biological Attach
Animal disease outbreak	Dam failure	Chemical attack
Drought	Levee failure	Cyber incident
Earthquake	Mine accident	Explosives attack
Epidemic	Hazardous materials release	Radiological attack
Flood	Power failure	Sabotage
Hurricane	Radiological release	School and workplace violence
Landslide	Train derailment	
Pandemic	Urban conflagration	
Tornado		
Tsunami		
Volcanic eruption		
Wildfire		
Winter storm		

CPG 201 discusses *Threat and Hazard* as the first step of defining the potential risk. Traditionally, NFPA documents start with fire or electricity as the threat or hazard. The broader application for community disaster resilience must address a wide range of initiating events. CPG 201 categorizes these potential event/threats as Natural, Technological, or Human-Caused.

CPG 201 outlines two factors for selecting the threats needing to be evaluated:
Factor #1 Likelihood:

Likelihood is the chance of something happening, whether defined, measured, or estimated objectively or subjectively. **Communities should consider only those threats and hazards that could plausibly occur.**

Factor #2 Significance of Effects:

The threat/hazard effects represent the overall impacts to the community. **Communities should consider only those threats and hazards that would have a significant effect on them**

3.3 Risk Informed Decision Making

Risk Informed Decision Making

Probability	Description
Frequent	Likely to occur frequently, experienced ($p > 0.1$)
Probable	Will occur several times during system life ($p > 0.001$)
Occassional	Unlikely to occur in a given system operation ($p > 10^{-6}$)
Remote	So improbable, may be assumed this hazard will not be experienced ($p < 10^{-6}$)
Improbable	Probability of occurrence not distinguishable from zero ($p \approx 0.0$)

Severity	Impact
Negligible	The impact of loss will be so minor that it would have no discernible effect on the facility or its operations.
Marginal	The loss will have impact on the facility, which may have to suspend some operations briefly. Some monetary investments may be necessary to restore the facility to full operations. Minor personal injury may be involved.
Critical	The loss will have a high impact on the facility, which may have to suspend operations. Significant monetary investments may be necessary to restore to full operations. Personal injury and possibly deaths may be involved.
Catastrophic	The fire will produce death or multiple deaths or injuries, or the impact on operations will be disastrous, resulting in long-term or permanent closing. The facility would cease to operate immediately after the fire occured.

Some analysis or quantification of the risks identified is necessary to allow comparison of options. A common starting point is the ranking of Risk based on likelihood ranges and severity categories. Chapter 5, Risk Criteria of the *Guidance Document for Incorporating Risk Concepts into NFPA Codes and Standards* and NFPA *551 Guide for the Evaluation of Fire Risk Assessments*, 2016 Edition use an example approach from MIL-STD-882E. This method uses five levels of likelihood and four categories of severity, as shown below.

The NIST's *CRPG* offers a similar approach using the term **Hazard Levels:**

For each hazard identified, communities are encouraged to determine three levels of the hazard for planning:

Routine—*Hazard level is below the expected (design) level and occurs more frequently. Resilient buildings and infrastructure systems should remain fully functional and not experience any significant damage that would disrupt social or economic functions in the community.*

Expected—*Design hazard level, where the design level is often based on codes. The design hazard level may be greater than the minimum required by codes, or may be based on other criteria. Buildings and infrastructure systems should remain functional at a level sufficient*

3.3 Risk Informed Decision Making

to support the response and recovery of the community as defined by the performance levels. This level is based on the design criteria normally used for buildings.

Extreme—Hazard level is above the expected (design) level. Some hazards refer to the maximum considered event, which is based on the historic record. Extreme events may also include long-term changes in hazards anticipated due to climate change. However, this hazard level might not be the largest possible hazard level that can be envisioned, but rather one that the community believes is credible. Critical facilities and infrastructure systems should remain at least minimally functional at this level. Other buildings and infrastructure systems should perform at a level that protects the occupants though they may need to be rescued. In addition, emergency response plans should be developed for scenarios based on this hazard level.

An earthquake example of these ***Hazard Levels*** is provided in the CRPG that referenced the design hazard levels in ASCE/SEI Standard 7-10 (ASCE/SEI 2010):

ROUTINE	EXPECTED	EXTREME
50 years	500 years	2,500 years

3.3.1 Decision Process

The decision process weighs the Risk—likelihood X consequence—against the risk metrics. This process implies there is an acceptable level of risk. Reaching a consensus on acceptability of risks is a major challenge. However, in comparing alternative strategies, the best use of resources will likely be the one that has the greatest influence on reducing the risk.

In current NFPA codes and standards, much emphasis is placed on preventing undesired events (fires, shock hazards, etc.). In the case of the natural threats identified in CPG 201, the initiating event cannot be prevented by any actions prescribed in codes and standards. Therefore, in the context of *Community Disaster Resilience* the risk reduction strategies will focus on withstanding, responding to, and recovering from these initiating events. For the human-caused threats and some technological threats, prevention measures may be appropriate, but not to the exclusions of withstanding, responding to, and recovering from these initiating events. In general terms, the decision process characterized the risk and determines if the risk metrics are acceptable. The process may become iterative, if the first (unmitigated or unchanged) characterization results in unacceptable metrics. Figure 3.1 illustrates the general process. In the case of codifying the resulting mitigation plan, it will be critical to clearly explain the risk metrics, leading to the performance objectives, resulting in the requirements.

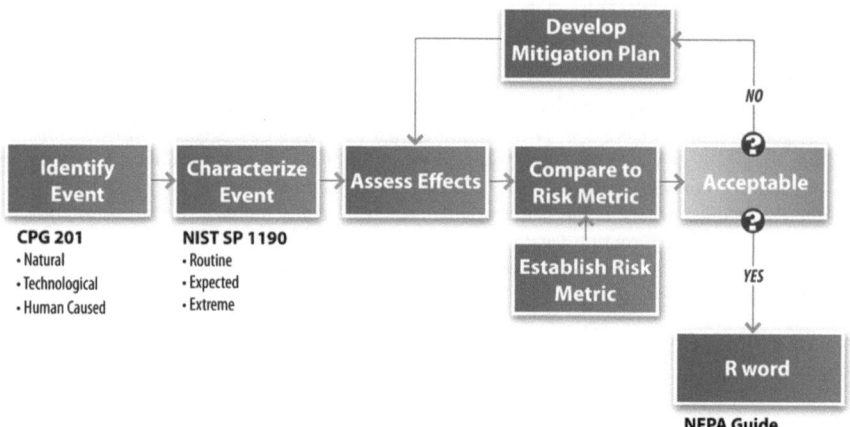

Fig. 3.1 Decision process

3.3.2 Structural Example

Risk Category of Buildings and Other Structures for Flood, Wind, Snow, Earthquake, and Ice Loads	
Use or Occupancy of Buildings and Structures	Risk Category
Buildings and other structures that represent a low risk to human life in the event of failure.	I
All buildings and other structures except those listed in Risk Categories I, III, and IV.	II
Buildings and other structures, the failure of which could pose a substantial risk to human life.	III
Buildings and other structures, not included in Risk Category IV, with potential to cause a substantial economic impact and/or mass disruption of day-to-day civilian life in the event of failure.	III
Buildings and other structures not included in Risk Category IV (including, but not limited to, facilities that manufacture, process, handle, store, use, or dispose of such substances as hazardous fuels, hazardous chemicals, hazardous waste, or explosives) containing toxic or explosive substances where their quantity exceeds a threshold quantity established by the authority having jurisdiction and is sufficient to pose a threat to the public if released.	III
Buildings and other structures designated as essential facilities.	IV
Buildings and other structures, the failure of which could pose a substantial hazard to the community.	IV
Buildings and other structures (including, but not limited to, facilities that manufacture, process, handle, store, use, or dispose of such substances as hazardous fuels, hazardous chemicals, or hazardous waste) containing sufficient quantities of highly toxic substances where the quantity exceeds a threshold quantity established by the authority having jurisdiction to be dangerous to the public if released and is sufficient to pose a threat to the public if released. [a]	IV
Buildings and other structures required to maintain the functionality of other Risk Category IV structures.	IV

[a] Buildings and other structures containing toxic, highly toxic, or explosive substances shall be eligible for classification to a lower Risk Category if it can be demonstrated to the satisfaction of the authority having jurisdiction by a hazard assessment as described in Section 1.5.2 that a release of the substances is commensurate with the risk associated with that Risk Category.
Source ASCE/SEI Standard 7-10 (ASCE/SEI 2010), Table 1.5-1

ASCE/SEI Standard 7-10 (ASCE/SEI 2010) and its incorporation into the building codes addresses risk informed design requirements by identifying the design loads a structure must support. The standard attempts to categorize risk in terms of the

3.3 Risk Informed Decision Making

size of the population exposed and the critical function of the facility. These categories are expressed in the table shown below. The drawback of this approach is the mixing of apparent risk metrics in the same category: population at risk from the event and need for the facility after the event. These categories are then used throughout the standard for establishing design loads.

In addition to the Risk Categories, the structural standard identifies *Importance Factors* for each category and each imposed load.

> *Minimum design loads for structures shall incorporate the applicable importance factors given in the following table, as required by other sections of this Standard.*

This table does not include component importance factor, I_p, applicable to earthquake loads, because it is dependent on the importance of the individual component rather than that of the building.

Importance Factors by Risk Category of Buildings and Other Structures for Snow, Ice, and Earthquake Loads

Risk Category from Table 1.5-1	Snow Importance Factor I_s	Ice Importance Factor-Thickness I_i	Ice Importance Factor-Wind I_w	Seismic Importance Factor I_e
I	0.80	0.80	1.00	1.00
II	1.00	1.00	1.00	1.00
III	1.10	1.25	1.00	1.25
IV	1.20	1.25	1.00	1.50

Source ASCE/SEI Standard 7-10 (ASCE/SEI 2010), Table 1.5-2

The importance factor provides a means of addressing the different loads (a.k.a. initiating events) for the different risk categories (a.k.a. risk metrics).

The building codes have expanded on the ASCE standard to clarify some of the risk metrics, in NFPA 5000 Table 35.3.1 and IBC Table 1604.5. These tables further define structures, *which could pose a substantial risk to human life*. Threshold values on occupant load are established for occupancies such as schools, colleges, assembly occupancies and health care. But facility importance is implied as well, since risking 500 college students in a structure is equated to risking 250 elementary school students.

Expanding the structural approach to a broader application for *Community Disaster Resilience* cannot be successful without identifying and perhaps separating the facility/building use before, during and after the event, from the occupant load or population at risk. This distinction will be essential in the response and recovery aspects of *Community Disaster Resilience*.

3.3.3 NIST Framework

NIST's *Community Resilience Planning Guide for Buildings and Infrastructure Systems* (CRPG) offers a performance-based slant on classifying structures and facilities. Instead of referencing facility type or occupant loading, it establishes four categories based on performance level in response to the initiating event. This approach is very useful in establishing performance requirements for supporting components and systems, such as fire safety features, in response to the threats selected in the risk identification process. The table below identifies four categories and the expected performance level of the structures, system and components.

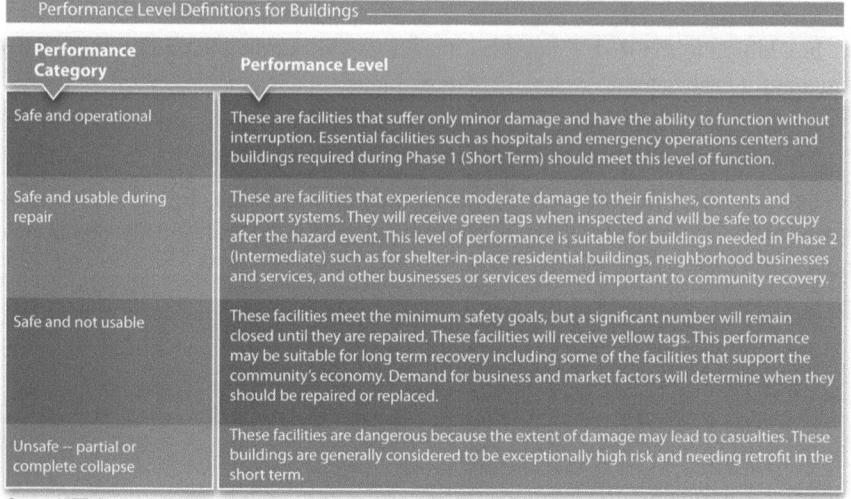

Source: NIST's Community Resilience Planning Guide for Buildings and Infrastructure Systems, Special Publication 1190, Table 4.1

In contrast to the Risk Categories expressed in ASCE/SEI Standard 7-10 (ASCE/SEI 2010) and expanded in NFPA 5000 Table 35.3.1, the NIST performance categories help establish priorities for response and recovery as well as design criteria.

3.3.4 Decision Trees

Another approach to decision support tools is the Decision Trees. An example of this tool is contained in NFPA 550, *Guide to the Fire Safety Concepts Tree*. The purpose of NFPA 550 is expressed in the Guide, paragraph 1.2:

> This guide is intended to assist the Fire Safety Practitioner (e.g., Designer, Engineer, Code Official) in communicating fire safety and protection concepts. Its use can assist with the analysis of codes or standards and facilitate the development of performance-based designs.

One section of the Tree applies equally well to *Community Disaster Resilience*: **"Manage Exposed."** As seen below, the concepts of *Defend Exposed in Place* or *Move Exposed*, applies to many if not all of the events/threats identified in Sect. 3.2. These concepts also fit into the framework of the CRPG and the building codes' structural standard. The top tier is similar to establishing the limits on the population at risk (e.g., 250 elementary school children, 500 college students, 5000 office workers) as implied in NFPA 5000, Table 35.3.1. The concepts for safeguarding the exposed expressed in the tree are useful reminders of what is necessary to achieve the goal. Simply stated, evacuation of an exposed population requires the ability to Cause Movement (communications), the Means of Movement, and a Safe Destination.

3.3 Risk Informed Decision Making

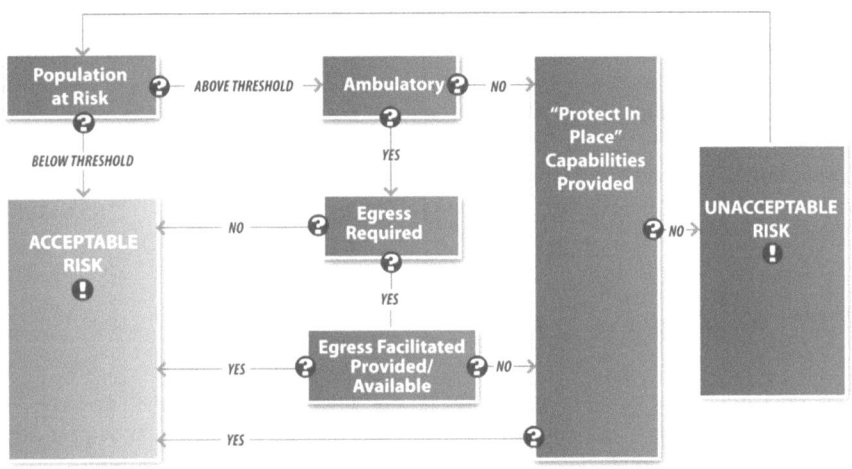

Fig. 3.2 Population at risk

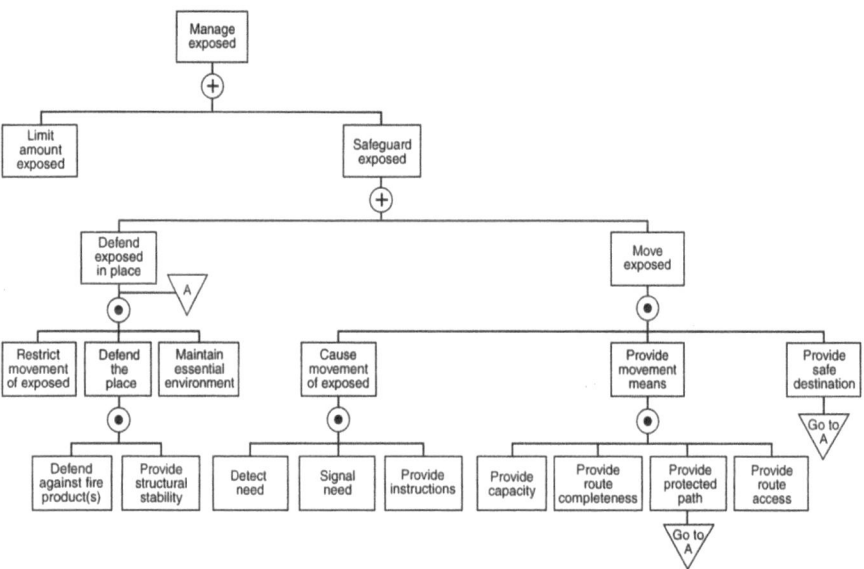

Source: NFPA 550, **Figure 4.5.2.1 "Manage Exposed" Branch of Fire Safety Concepts Tree.**

Figure 3.2 presents an example flow diagram for decisions related to a key performance objective of emergency response as expressed in PPD 8, *National Preparedness*:

The term "response" refers to those capabilities necessary to save lives, protect property and the environment, and meet basic human needs after an incident has occurred.

The capability to save lives and meet basic needs is reflected in the diagram. The elements in the *Fire Safety Concept Tree* from NFPA 550 and those in Fig. 3.2 work in concert. If evacuation and relocation is necessary, communication, transportation (means and routes) and a *safe* destination must be provided. The bases for each of the decisions can be expanded for **Engineered Features** and **Administrative/Operational Features** available or desired. If evacuation capabilities are not provided or available, the exposed location must be made *safe*.

3.3.5 Resulting Criteria

The decision processes described above are intended to result in criteria for **Engineered Features** and **Administrative/Operational Features** that support the *community disaster resilience* goals. The inclusion of these criteria in NFPA documents introduces the question of form. Should these criteria be mandatory in enforceable language? Should these criteria be in guides and recommended practices? Should guidance documents focus on process methodologies rather than prescriptive requirements? Technical Committees that wish to address the resiliency concepts discussed above must determine what best serves the mission of the NFPA. The gathering and sharing of knowledge and information can take many forms.

Chapter 4
Buildings and Infrastructure

Technical Committee member considering incorporating concepts of disaster resilience are encouraged to study the NIST *Community Resilience Planning Guide for Buildings and Infrastructure Systems, Special Publication 1190*. The performance categories provide an essential context for considering the role of NFPA documents. For the purposes of this Chapter, the discussion will be divided into two main areas: existing NFPA codes and standards and future NFPA documents.

4.1 Existing NFPA Codes and Standards

This section addresses the existing scopes of NFPA codes and standards the primarily address **Engineered Features** in the built environment. The scopes of these documents typically address fire as the hazard and purport to establish minimum requirements to mitigate the hazards. So the natural questions become:

1. What is the performance objective of the prescribed mitigating feature(s) in the event of any of the Threat event identified in CPG 201?
2. Are the features prescribed *resilient* with regard to those threats?

The performance objectives served by the prescriptive requirements contained in the code or standard must be understood. The validity of those objectives during and after the external threat must be assessed. The starting point is to evaluate the purpose or role of the facility to which the NFPA document can be applied. The NFPA requirements may be based on the occupancy or hazard. These requirements anticipate the normal range of uses for the buildings and systems, without the introduction of any external threat. It is recognized that this is a generalization and that some NFPA document contain some requirements relating to non-fire events. However, to address the occupancy or use issue in concert with any external threat, the performance category and performance levels contained in the CRPG should be applied. The NIST performance levels must be matched with the risk metrics implied or

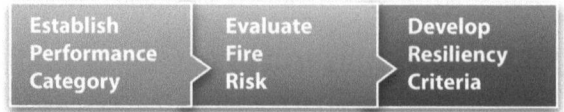

Fig. 4.1 Three step process

explicit in the NFPA document. For example, the requirement for sprinkler protection of an assembly occupancy that was based on an occupant load ostensible is based on safety to life. If the facility is empty, the sprinkler system mission becomes property protection alone. If however, that facility is chosen in the community for an emergency shelter, the life safety mission of the sprinkler system remains valid. The four stages of use to be considered are:

- Normal Conditions
- During Event
- Post Event Recovery Period
- Restoration

Normal conditions typically establish the performance expectations for the design features. The use/purpose/role of the facility may change during the external event. So those performance expectations may change. Likewise, after the event, the timeline for recovery and restoration may create time dependent variations on performance expectations.

Figure 4.1 shows the three step process for determining what, if any, changes should be recommended for added resilience.

4.1.1 Step 1: Establish Performance Category

It should be determined if the performance categories outline by NIST are consistently applied across all communities or NFPA document users. If the categories are embraced, the guidance provided in NFPA documents could start with Category A, or Category B as starting points. These categories establish rationale for discriminating between facilities and allow optimizing performance.

4.1.2 Step 2: Establish Fire Risk

The fire safety features contained in NFPA documents are based on a perceived level of fire risk. Using the risk metrics referenced in Sect. 3.1, the change in fire risk brought about by the external event/threat can be evaluated. Does the event increase the likelihood of an ignition? Can the event actually start a fire

4.1 Existing NFPA Codes and Standards   21

(lightning, wild fire, gas line break, electrical fault, etc.)? Or does the event reduce the potential for ignition (flood, power outage, severe cold, etc.)? Does any change in use of the building impact the ignition potential? In summary the likelihood of a fire occurring could go up, remain the same, or go down. Regarding consequences or magnitude of potential fires, three separate metrics could be considered. First, does the population at risk change? If the building is evacuated and remains so until restoration, the population at risk drops as the building is evacuated. If however, the building assumes an additional role during and/or after the event, that role could increases the population at risk above the normal use. If the facility is a shelter in place, the population at risk could remain unchanged. The second is the risk of property damage. If fire safety features are impaired by the event, fire growth and spread unchecked could increase property damage. The third is the continuity of operation. If the operations of a facility are critical to emergency response, the consequence of it interruption could increase during and/or after the event.

Figure 4.2 displays this evaluation of the potential change in fire risk. A useful screening tool for Step 2 could be a Risk Matrix as described in NFPA 551. Figure 4.3 illustrates the relative fire risks matrix for evaluating changes in risk resulting from an event/threat. The consequence metric could be any of those express above.

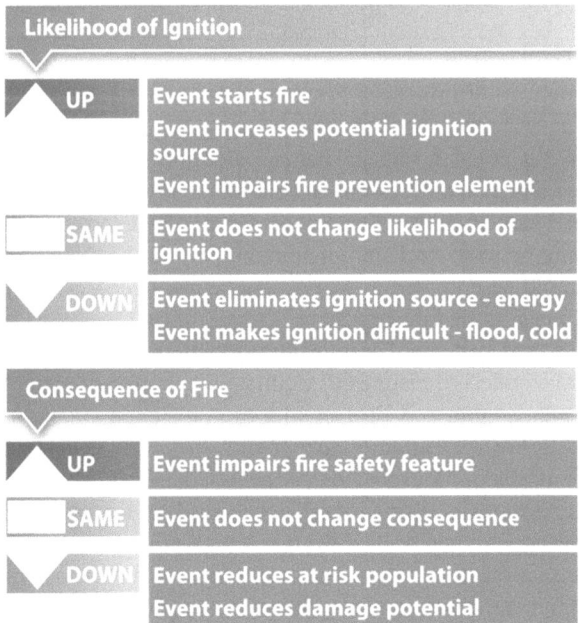

Fig. 4.2 Evaluate fire risk

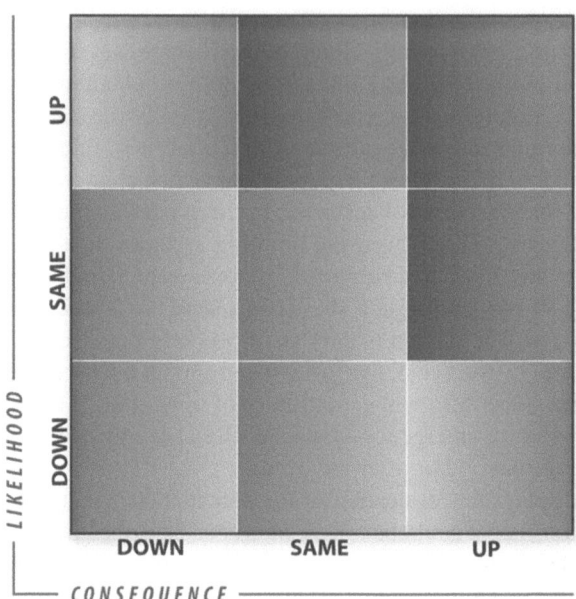

Fig. 4.3 Risk matrix

4.1.3 Step 3 Establish Resilience Criteria

Considering the key works explained in Sect. 1.2, Resilience Criteria will most likely include W*ithstanding disruptions* and *Recovering rapidly from disruptions*. Since this section focuses on **Engineered Features**, *Withstanding disruptions* could mean robust, redundant, reliable design changes to ensure urgently required fire protection features remain functional during and perhaps after the event/threat. The cost-benefit of any proposed changes will depend on the risk identification outlined in Sect. 3.2. Because of the wide variability of event/threat risk, the determination of *if* changes are required may best left on local jurisdiction, but the *how* the changes could be implemented could be provided in NFPA documents. *Recovering rapidly from disruptions* could mean similar design changes to minimize time for repair and restoration of structures and systems. These recommendations could be design changes to minimize damage or design features to facilitate easy repair or replacement.

In Parallel with **Engineered Features** for fire protection system resiliency, those features related to fire prevention in *hazard* and *occupancy* documents may also be revisited for the impact of the events/threats identified in Sect. 3.2. If the first two steps identified above show the initiating event/threat capable of increasing the likelihood of a fire, resiliency criteria could be developed to mitigate those conditions. One example is an automatic shut off for a fuel source (natural gas or oil). Another example could be a shut off for a process or potential ignition source (furnace, space heater, etc.).

4.2 Future NFPA Documents

The success of the voluntary consensus standards process is its ability to bring together the correct mixture of stakeholders and technical expertise. For the process to work, the standards writing organizations must collaborate and not compete. Those **Engineered Features** that are a natural extension of the current NFPA Technical Committee activities should be considered if those voids are not being filled by other organizations (regulators, codes bodies or consensus standards organizations). If an expansion of committee scope is necessary, approval of the Standards Council is required prior to the expenditure of committee resources.

Chapter 5
Emergency Preparedness—Planning, Response and Recovery

This section is intended to address those aspects of *Community Disaster Resilience* that go beyond the **Engineered Features** discussed in Chap. 4. A valuable starting place is to repeat the National Preparedness Goals measures of success:

> *A secure and resilient nation with the capabilities required across the whole community to prevent, protect against, mitigate, respond to, and recover from the threats and hazards that pose the greatest risk.*

The National Preparedness Goals as discussed in Sect. 3.1 cover five missions included in the above quote. Many aspects of *mitigate, respond to and recover from* require **Administrative/Operational Features** when **Engineered Features** fail or are inappropriate. The challenge of these plans and activities is *the ability to prepare for and adapt to changing conditions*, as highlighted in PPD21. Adapting to changing conditions is perhaps better served by outlining decision making processes than prescriptive requirements.

5.1 Defining Community

Both NIST's CRPG and FEMA's NDRF emphasize the connectedness within and between communities necessary for *Disaster Resilience*. It is unrealistic to expect that a connected and collaborative response to and recovery from an event/threat will occur without adequate planning. Perhaps the first step of that planning is defining the community capable of planning and taking action. The magnitude of the event/threat will influence how broadly the community is defined (e.g. town, County, State, region, Nation). Each community has a vital role in planning for its *Disaster Resilience*. Chapter 10 of the NDRF discusses some of the considerations related to defining and coordinating community planning and response to event/threats.

5.2 Achieving Community Disaster Resilience

Using the PPD21 definition of resilience as, *the ability to prepare for and adapt to changing conditions and withstand and recover rapidly from disruptions,* it is clear that achieving *Community Disaster Resilience* will requires **Planning, Resources and Practice**. Those activities and decisions will vary from community to community and from event/threat to event/threat. Chapter 5 of the NDRF provides insight into achieving these resiliency goals in terms of recovery:

> Each community defines successful recovery outcomes differently based on its circumstances, challenges, recovery vision and priorities. One community may characterize success as the return of its economy to pre-disaster conditions while another may see success as the opening of new economic opportunities. Although no single definition fits all situations, successful recoveries do share conditions in which:

- *The community successfully overcomes the physical, emotional and environmental impacts of the disaster.*
- *It reestablishes an economic and social base that instills confidence in the community members and businesses regarding community viability.*
- *It rebuilds by integrating the functional needs of all residents and reducing its vulnerability to all hazards facing it.*
- *The entire community demonstrates a capability to be prepared, responsive, and resilient in dealing with the consequences of disasters.*

The capacity to be prepared and responsive requires **Planning, Resources and Practice.**

5.2.1 Planning

One common thread through all the DHS, NIST and NFPA documents is the importance of **Planning.** A well-organized explanation of the goals and objectives of disaster recovery planning is contained in Chap. 9 of the NDRF. The *resiliency* afforded by this planning is a rapid and successful recovery:

> Preparedness initiatives help guide the recovery process to effectively and efficiently achieve a community's disaster recovery priorities. Both pre- and post-disaster recovery planning are critical for communities to develop resilience and for successful and timely recovery.

It is important to note that the planning advocated by the NDRF is both prior to an event/threat—Pre-disaster planning—and also after an event/threat—Post-disaster planning. NDRF Chap. 9 provides Key Elements and Recommended Activities for both Pre-Disaster and Post-Disaster. The Risk-Informed process described in Sect. 3.3.1 above, can be applied to this planning process as well. The *Assessment Activities* (identification, characterization and damage assessment) and the *Guiding Principles and Recovery Priorities* (risk metrics and analysis) lead to mitigation and recovery plans that are responsive to the needs and resources of the community.

5.2 Achieving Community Disaster Resilience

NFPA 1600, *Standard on Disaster/Emergency Management and Business Continuity Programs*, mirrors many of the elements discussed in CPG 201 and the NDRF. The enforcement language of *shall* requirements is worth noting. The standard does not establish that a *program* shall be established, but in its Sect. 1.1 Scope states:

> This standard shall establish a common set of criteria for all hazards disaster/emergency management and business continuity programs....

Specific guidance regarding decision processes, acceptability of risks, and resource allocation priorities is difficult and perhaps undesirable to prescribe in mandatory language. Likewise, the adaptability required for resiliency as outline above in Sect. 1.2 does not lend itself to prescriptive requirements. However, there is a need for guidance on the planning processes to help all aspect of the *community* understand and participate in the establishment of priorities. The success of the planning and implementation for *community disaster resilience,* as clearly stated in the NIPP 2013 for our infrastructure, requires cooperation and coordination between private and public organization and among effected populations. Determining the type and details of guidance, necessary to foster this cooperation, is a worthy challenge. Determining the appropriate role of NFPA is also an appropriated exercise.

As referenced in Sect. 3.3.4, Decision Trees or Concept Trees can be useful planning tools. In the example in Sect. 3.3.4, the decision process for resources and activities necessary to mitigate exposure to the event/threat can be supported, using the "Manage Exposed" tree from NFPA 550. If the portion of the population in harm's way is not "protected in place" the features necessary to move them are identified: Notification/communications; Means of movement (means of egress, means of transportation, transportation routes, etc.) and Safe destination (shelter, high ground, etc.)

Planning for Post-disaster recovery will also require risk-based decisions. The *Recovery Continuum* helps identify the timeliness of restoring basic health and safety functions. The NIST CDRF in its Chap. 11 discusses "functionality goals" for different levels of performance. With regard to facility reuse or reentry, the total risk to the health and wellbeing of an exposed population must be weighed in deciding the level of impairment of fire safety features that can be tolerated. Occupying or sheltering in a less that completely restored facility may be less risk to the health and wellbeing of the exposed population than being without use of the facility. These reuse decisions must be based on risk metrics and methods discussed in Chap. 3. Short term (as illustrated in the *Recovery Continuum* in Appendix 2) exposure to an increased risk of fire may be far less than the health and wellbeing risk of not reusing the facility. The methodology for making these "reuse" decisions must be included in the post-disaster planning.

5.2.2 Resources

Successful response to event/threats requires the effective and efficient deployment of available resources. Those resources will include personnel, equipment, materials and infrastructure (facilities). Resource allocation is integral with the planning.

The planning process can often result in a "wish list" of all the things desired to implement the plan. In these cases the plan drives the resource needs. It is sometimes more effective to start with the resources available to determine the most effective ways to deploy those resources. Section 5.4 of NFPA 1600 addresses a Resource Needs Assessment. This section presents important concepts for determining the resources needed, and how and when those resources can be accessed.

Personnel: The people exposed to and responding to the event/threat can be the greatest asset in community disaster resilience. The roles and responsibilities cataloged in Appendix B of the NDRF provide an excellent road map to making these human resources assets rather than liabilities. Therefore, one important aspect of resource planning is public engagement and awareness. Although these types of activities are not NFPA codes and standards subjects, they are well within the mission and expertise of NFPA.

Most often the first responders depended on for these event/threats are the fire fighters and EMTs. Standards such as NFPA 1201, *Standard for Providing Fire and Emergency Services to the Public*, NFPA 1710, *Standard for the Organization and Deployment of Fire Suppression Operations, Emergency Medical Operation, and Special Operations to the Public by Career Fire Departments,* and NFPA 1720, *Standard for the Organization and Deployment of Fire Suppression Operations, Emergency Medical Operation, and Special Operations to the Public by Volunteer Fire Departments* establish the expectation for "routine" response. Annex B in both NFPA 1710 and 1720 provides a model for community wide risk management with a focus on fire safety. Figure B.1.2 *Fire Safety Concepts for Fire Department Operations* is similar in approach as NFPA 550's Fire Safety Concepts. To support these existing organizations in playing an effective role in disaster response and recovery, similar models could be developed in concert with the FEMA's NDRF and NIST's CDRF that identify other *Community Disaster Resilience* concepts.

Likewise, NFPA produces many standards regarding the qualifications necessary to perform fire fighter anticipated jobs. Disaster response and recovery may require new qualifications. The consensus process may be the appropriate approach to cataloging those qualifications. A good example of this is NFPA 1006, *Standard for Technical Rescuer Professional Qualifications.* However, these additional qualifications may need to be applied to other resources in addition to or in place of the fire fighters.

Successful disaster response and recovery will likely exploit resources that are not professionals, especially when the event/threat overwhelms the typically small pool of professional responders. The safe and effective application of these volunteer resources presents many planning and management challenges, where consensus guidance could be useful. The concept of Community Emergency Response Teams (CERT) has been promoted by FEMA to help improve response to event/threats and exceed the immediate response capabilities of the normal emergency responders. Organized, trained and deployed properly, these teams can have a significant positive impact.

Equipment: NFPA standards exist for the design and equipping of emergency vehicles e.g. NFPA 1901, NFPA 1906). Developing minimum standards surrounding the safety and performance of these vehicles is appropriate. In some cases, standards developed by different organizations may exist for the same equipment.

Insurance Services Office, Inc (ISO) has minimum equipment lists for pumpers, ladder trucks, and service companies as part of their basis for municipal protection class grading. NFPA also has these lists. While it is clear that various hose and tools are required to perform firefighting and other emergency response tasks, it seems that the type and number of these various types of equipment should be based on the performance objectives defined for the first responders and consistent with the outcomes designed by the community stakeholders. What is the purpose of requiring specific lengths of supply hose and attack lines if hydrant spacing in the community is either much shorter or longer than the required lengths? Many departments design hose loads to match their community needs. It may be more practical to prepare a recommended practice to ensure that the compliment of hose and equipment carried by the fire department's vehicles collectively are capable of providing the performance measures that support the community's desired outcomes.

Equipping the first responder fleet for a disaster response is another matter. The community has to define the types of potential disasters and then whether the response to the event requires the community to be self-sustaining or can expect regional, state, or federal assets. The relationship between simply sustaining the government infrastructure and investing responder resources in assisting key business interests (such as tourist attractions, large employers) that drive the tax base must be defined. Current standards in many cases focus on prescriptive lists of personnel, skills, equipment, and vehicles, but very little on what should be accomplished by having all of these resources. It is a lot of thou shall have standards (inputs) rather than thou shall insure that the following outcomes are achieved. For the day to day, higher frequency/low consequence events, this is relevant to define, but not likely to impact the resiliency of the community. However, the low frequency/high consequence events are potentially community impacting and are exactly the kinds of events that should be addressed in terms of resiliency. The extent of natural events such as tornados, hurricanes, and earthquakes can be modeled and past experiences studied. Maximum events of these kinds are rarely, if ever, handled with only with local community resources. If fact, the local resources may be destroyed. It is more appropriate from an equipment perspective to maintain resources capable of handling the relevant higher frequency/low consequence events and defer to the planning component to have access to needed equipment from the higher consequence events.

Materials: The materials required to respond to event/threats identified will include a vast array of items from building materials, to fuels, to health and hygiene supplies, to food and water. NFPA 1600 in paragraph 5.4.3 addresses establishing procedures to *locate, acquire, store, distribute, maintain, test, and account for services, human resources, equipment and material....* The requirements in the standard are expanded on in annex material, A.5.4.3. The list in A.5.4.3 is a good starting point for cataloging both needs and resources. The key words in paragraph 5.4.3 also address actions: *locate, acquire, store, distribute, maintain, test, and account for.* A challenge for the community in the materials aspect of the planning is quantifying the materials needed, the material available and the priorities for sharing those materials. The *Recovery Continuum* in the NDRF is useful in establishing those priorities and how those priorities could change with time during post-disaster planning.

5.2.3 Practice

As part of the commissioning and acceptance of a building or system, testing is performed to verify things work as intended (or at least as designed). The same consideration should be applied to the emergency preparedness programs at all levels. Training on and exercising of the program elements is essential to demonstrate expectations can be met. Likewise, as periodic testing, inspection and maintenance is required for buildings, systems and equipment, so are periodic reviews and exercises are necessary for response and recovery plans. Individual elements of the programs and plans and be exercised, but the coordination and collaboration of the various elements and organization needs to be tested as well.

Training: A starting point is the training of individuals and groups on the skill sets and on the roles and responsibilities necessary to implement the plan. As stated above under personnel, required qualifications can be established. In conjunction with the establishment of qualification, the training requirements necessary to acquire those qualifications (skills and understanding) need to be identified. This training can be job related as for emergency responders on public awareness related for the vast array of volunteers. The consensus process could be an effective approach to identifying training requirements and methods. Likewise public and community awareness activities could be guided by consensus performance goals.

FEMA's *National Incident Management System (NIMS)—Training Program*, 2011 provides a framework for training to support incident management:

> *The NIMS Training Program lays out a conceptual framework that maintains a systematic process for the development of training courses and personnel qualifications. This process produces trained and qualified emergency management personnel. The framework facilitates the systematic development of these courses and qualifications by translating functional capabilities (defined in NIMS) into positions, core competencies, training, and personnel qualifications. The NIMS Training Program sets a sequence of goals, objectives, and action items for the NIC, which administers NIMS training nationally, and for stakeholders, who run their respective NIMS training and education programs.*

The process developed and implemented by FEMA address the importance management aspect of disaster response. These training elements are critical to the response to and recovery from disasters. FEMA defines the strategic objectives of the *NIMS Training Program* as follows:

Defines a national curriculum for NIMS and provides information on NIMS courses in the core curriculum, applicable to all levels of government, the private sector, and NGOs by promoting comprehensive NIMS-related training beyond Incident Command System (ICS) training.

Identifies broad NIMS training goals and objectives for NIMS national training for both the NIC and stakeholders and outlines guidance to attain them.

Guides human resource management via established training baselines for emergency and incident response personnel qualifications, based on development of core competencies for NIMS-based incident management positions.

Conveys information pertaining to instruction and learning, articulating specifications for the consistent delivery of NIMS training through a national baseline curriculum for NIMS, with each course having objectives that meet training needs set by the core competencies, complete training guidance, and instructor qualification guidelines.

Guides development of stakeholders' long-term training plans, budgets, and schedules as well as grant qualifications and applications.

The *NIMS Training Program* provides an approach that establishes a baseline training requirements, then adds additional levels of training intended to prepare emergency managers for more complex incidents. These courses are intended to develop management skill, rather than individual responder skills.

FEMA has also developed a program for Community Emergency Response Teams (CERT). In the *Community Emergency Response Team Basis Training Participants Manual* the supportive role of the CERT is defined:

> If available, emergency services personnel are the best trained and equipped to handle emergencies. Following a catastrophic disaster, however, you and the community may be on your own for a period of time because of the size of the area affected, lost communications, and unpassable roads.

> CERT Basic Training is designed to prepare you to help yourself and to help others in the event of a catastrophic disaster. Because emergency services personnel will not be able to help everyone immediately, you can make a difference by using your CERT training to save lives and protect property.

The CERT approach can be a valuable asset to community resilience, because they engage the populace in the response and recovery. This engagement leads to a stronger sense of community and greater collaboration.

The premise that emergency services personnel are best trained leads to the question of what is the appropriate level of training for event/ threats identified in the communities risk assessment. One approach for individual responders could be that NFPA redefine a baseline responder (e.g. fire fighter) training standard then layer additional specialties. A corollary to this is found in the hazardous materials standard, NFPA 472, *Standard for Competence of Responders to Hazardous Materials/ Weapons of Mass Destruction Incidents*. The standard has a baseline for hot zone work defined in the hazardous materials technician. Then there are seven areas of specializing. The tie in to the resiliency concept is that all communities and the outcomes they are planning for are not the same. Nor are the event/threats they may identify. Some baseline requirements would need to be supplemented with a menu specifies based on the community needs. It would be appropriate to provide guidance on how to define desired outcomes and then measure the outcomes that a community desires to achieve. Then the potential resources available could be trained to achieve the desired outcomes.

Exercises: Emergency response is a team activity. As such the only way to evaluate the collaboration and cooperation aspects is to perform exercises which simulate all aspect of the plan. These exercises also become a valuable training tool for reinforcing role and responsibilities. NFPA 1600, *Chap. 8, Exercises and Tests,* addresses this evaluation and training tool. The scope and frequency of exercising the emergency response plan will depend on the complexity of the plan. Annex material A.8.5 recommends an annual exercise and testing of the emergency plan.

Smaller scope exercises or Drills can also be useful training tools for the "team". FEMA recommends drills for the CERT program. The FEMA website explains the resources available to assist communities to practice response:

> The National CERT Program has developed a library of drills and exercises. These exercises have been designed in a ready for-use format and include complete instructions, detailed lists of materials, and all supporting forms.

5.3 Recovery

The process of returning to normalcy after a significant event/threat will require the same pre-planning and decision making as the immediate response. The NDRF clarifies this point with the expressed timeline associated with the *Recovery Continuum*. As displayed in Fig. 1.3, the recovery process will require adaptability of the plans, decisions and actions to the changing conditions. The NIST CRPG concept of building/infrastructure functionality likewise implies the requirements for adaptability. As a corollary to the discussion in Chap. 4 above, the continuing reassessment of the risks of fire will be an essential element for the planning and implementing the recovery process.

In order to move toward recovery, some facilities may need to be reused before completely restored. Therefore, **Administrative/Operational Features** may be required before **Engineered Features** are completely restored. As stated above, if the health, safety and wellbeing risk to the potential occupants is greater without the reuse of the facility, than the short term risk of fire, the facility should be reoccupied. How that decision is arrived upon is an important element of the pre-planning process. One role for **Administrative/Operational Features** could be actions taken to compensate for the impairment to fire safety features. The measure, referred to as *interim compensatory measure,* in some regulatory arenas, can be attempts to reduce the likelihood of ignition by curtailing activities such as smoking, open flames or other hot surfaces. Another companion approach is to supplement or replace the fire safety impairment with human intervention, the most common of which is the use of trained "fire watches." Currently very little guidance is available on making those decisions on reuse and occupancy. The CRPG and NDRF provide a framework around which such guidance could be structured. A consensus approach to developing both a decision methodology and implementation guidelines could be a valuable addition to the disaster recovery process.

Appendix 1
Literature Review

The literature review is divided into sections, starting with definitions of resilience. As will be noted, there are a variety of definitions for resiliency or resilience, since the concept(s) has been applied very broadly. This review starts with wide definitions and then narrows to *disaster resiliency* as characterized in several federal government initiatives. Following the definitions is a discussion of the performance goal described in the literature as necessary to achieve resilience. Next is a review of suggested frameworks within which the codes and standards may be required to merge. Then there is a discussion of the role of codes and standards as defined by other research efforts in the literature.

Definitions of Resilience

A valuable starting place for understanding the evolution of the term and concepts of resilience is an article by Norris, F.P., et al. titled *Community Resilience as a Metaphor, Theory, Set of Capacities and Strategy for Disaster Readiness*. The article outlines 40 years of use of the term in physical, ecological, social, and community resilience. The article includes an extensive bibliography outlining this history. Norris defines resilience as *a process linking a set of adaptive capacities to a positive trajectory of functioning and adaptation after a disturbance.* That definition probably has more traction with community psychologists. However, it does raise two interesting concepts that (1) resilience is a process not just a result or outcome and (2) resilience is adaptive (dynamic not static). Zolli and Healy provide a valuable insight into the concept of resilience and how it could be applied to NFPA's mission and standards:

> Around the world, in disciplines as seemingly disconnected as economics, ecology, political science, cognitive science, and digital networking, scientists, policymakers, technologists, corporate leaders and activists alike are asking the same basic questions: What causes one system to break and another to rebound? How much change can a system absorb and still

> retain its integrity and purpose? What characteristics make a system adaptive to change? In an age of constant disruptions, how do we build in better shock absorbers for ourselves, our communities, companies, economies, societies and the planet?
>
> *Defining resilience more precisely is complicated by the fact that different fields use the term to mean slightly different things. In engineering, resilience generally refers to the degree to which a structure like a bridge or building can return to a baseline state after being disturbed. In emergency response, it suggests the speed with which critical systems can be restored after an earthquake or a flood…Though different in emphasis, each of these definitions rests on one of two aspects of resilience: continuity and recovery in the face of change.*

Although the definitions are diverse, the application to NFPA codes and standards will fall into one or both of these two aspects: *continuity or recovery*.

Another excellent reference with broad discussions on resilience is *Resilience Engineering: Concepts and Precepts*, edited by Hollnagel, Woods, and Leveson. This compilation of papers from a symposium offers insight into resilience engineering as a companion to safety management, both for systems and for organizations. One point related here from Westrum's discussion on *Resilience Typology* is particularly useful in understanding how codes and standards could influence resilience:

Protecting the organization from trouble can occur proactively, concurrently, or as a response to something that has already happened. These are all part of resilience, but they are not the same. Resilience thus has three major meanings.

- *Resilience is the ability to prevent something bad from happening.*
- *Or the ability to prevent something bad from becoming worse,*
- *Or the ability to recover from something bad once it has happened.*

This insight adds *prevention* either as part of, or in addition to, *continuity*.

The first official US government use of the concepts of resilience appears in the National Security Strategy, May 2010:

> At home, the United States is pursuing a strategy capable of meeting the full range of threats and hazards to our communities. These threats and hazards include terrorism, natural disasters, large-scale cyber-attacks, and pandemics. As we do everything within our power to prevent these dangers, we also recognize that we will not be able to deter or prevent every single threat. That is why we must also enhance our resilience—the ability to adapt to changing conditions and prepare for, withstand, and rapidly recover from disruption…

Improve Resilience Through Increased Public-Private Partnerships

> When incidents occur, we must show resilience by maintaining critical operations and functions, returning to our normal life, and learning from disasters so that their lessons can be translated into pragmatic changes when necessary. The private sector, which owns and operates most of the nation's critical infrastructure, plays a vital role in preparing for and recovering from disasters. We must, therefore, strengthen public-private partnerships by developing incentives for government and the private sector to design structures and systems that can withstand disruptions and mitigate associated consequences, ensure redundant systems where necessary to maintain the ability to operate, decentralize critical operations to reduce our vulnerability to

Appendix 1 Literature Review

single points of disruption, develop and test continuity plans to ensure the ability to restore critical capabilities, and invest in improvements and maintenance of existing infrastructure.

This policy use is expanded on in Presidential Policy Directive (PPD) 8, *National Preparedness*, which includes the following definitions of interconnected terms:
For the purposes of this directive:

(a) *The term "national preparedness" refers to the actions taken to plan, organize, equip, train, and exercise to build and sustain the capabilities necessary to prevent, protect against, mitigate the effects of, respond to, and recover from those threats that pose the greatest risk to the security of the Nation.*
(b) *The term "security" refers to the protection of the Nation and its people, vital interests, and way of life.*
(c) *The term "resilience" refers to the ability to adapt to changing conditions and withstand and rapidly recover from disruption due to emergencies.*
(d) *A threatened or actual act of terrorism. Prevention capabilities include, but are not limited to, information sharing and warning; domestic counterterrorism; and preventing the acquisition or use of weapons of mass destruction (WMD). For purposes of the prevention framework called for in this directive, the term "prevention" refers to preventing imminent threats.*
(e) *The term "protection" refers to those capabilities necessary to secure the homeland against acts of terrorism and manmade or natural disasters. Protection capabilities include, but are not limited to, defense against WMD threats; defense of agriculture and food; critical infrastructure protection; protection of key leadership and events; border security; maritime security; transportation security; immigration security; and cybersecurity.*
(f) *The term "mitigation" refers to those capabilities necessary to reduce loss of life and property by lessening the impact of disasters. Mitigation capabilities include, but are not limited to, community-wide risk reduction projects; efforts to improve the resilience of critical infrastructure and key resource lifelines; risk reduction for specific vulnerabilities from natural hazards or acts of terrorism; and initiatives to reduce future risks after a disaster has occurred.*
(g) *The term "response" refers to those capabilities necessary to save lives, protect property and the environment, and meet basic human needs after an incident has occurred.*
(h) *The term "recovery" refers to those capabilities necessary to assist communities affected by an incident to recover effectively, including, but not limited to, rebuilding infrastructure systems; providing adequate interim and long-term housing for survivors; restoring health, social, and community services; promoting economic development; and restoring natural and cultural resources.*

Likewise, PPD 21, *Critical Infrastructure Security and Resilience* expands the definition of resilience and charges the Department of Homeland Security with championing the national effort.

> *The term "resilience" means the ability to prepare for and adapt to changing conditions and withstand and recover rapidly from disruptions. Resilience includes the ability to withstand and recover from deliberate attacks, accidents, or naturally occurring threats or incidents.*
>
> *The Secretary of Homeland Security shall provide strategic guidance, promote a national unity of effort, and coordinate the overall Federal effort to promote the security and resilience of the Nation's critical infrastructure. In carrying out the responsibilities assigned in the Homeland Security Act of 2002, as amended, the Secretary of Homeland Security evaluates national capabilities, opportunities, and challenges in protecting critical infrastructure; analyzes threats to, vulnerabilities of, and potential consequences from all hazards on critical infrastructure; identifies security and resilience functions that are necessary for effective public-private engagement with all critical infrastructure sectors; develops a national plan and metrics, in coordination with SSAs and other critical infrastructure partners; integrates and coordinates Federal cross-sector security and resilience activities; identifies and analyzes key interdependencies among critical infrastructure sectors; and reports on the effectiveness of national efforts to strengthen the Nation's security and resilience posture for critical infrastructure.*

The PPD applications of the term *resilience* emphasizes the continuity aspect using withstand, the recovery aspect using recover, and the adaptive aspect using adapt. Later, the role of planning, or *prepare for,* will be discussed, more as a preparedness tool than a definition of resilience.

The pairing of the concept of *resiliency* with the characterization of the disruptive or adverse event as a disaster is well expressed by the National Academies' Committee on Increasing National Resilience to Hazards and Disasters. Their report *Disaster Resilience: A National Imperative* fine tunes the definition as:

> **Resilience**: *The ability to prepare and plan for, absorb, recover from, or more successfully adapt to actual or potential adverse events.*

The Committee further elaborates on the concept in its comment:

What Is Resilience?

> *Although resilience with respect to hazards and disasters has been part of the research literature for decades the term first gained currency among national governments in 2005 with the adoption of The Hyogo Framework for Action by 168 members of the United Nations to ensure that reducing risks to disasters and building resilience to disasters became priorities for governments and local communities (UNISDR, 2007). The literature has since grown with new definitions of resilience and the entities or systems to which resilience refers (e.g., ecological systems, infrastructure, individuals, economic systems, communities). Disaster resilience has been described as a Process, an outcome, or both, and as a term that can embrace inputs from engineering and the physical, social, and economic sciences.*

In response to PPD-21, Department of Homeland Security (DHS) embarked on two parallel efforts, *National Infrastructure Protection Program (NIPP) 2013, Partnering for Critical Infrastructure Security and Resilience* (initiated with NIPP 2006 and 2009) and *National Disaster Recovery Framework: Strengthening Disaster Recovery for the Nation,* developed by FEMA in 2011. NIPP 2013 builds on the definition of PPD-21 and defines resilience and recovery as follows:

Resilience—*The ability to prepare for and adapt to changing conditions and withstand and recover rapidly from disruptions; includes the ability to withstand and recover from deliberate attacks, accidents, or naturally occurring threats or incidents.*

Recovery—*Those capabilities necessary to assist communities affected by an incident to recover effectively, including, but not limited to, rebuilding infrastructure systems; providing adequate interim and long-term housing for survivors; restoring health, social, and community services; promoting economic development; and restoring natural and cultural resources.*

NIPP 2013 focuses on Critical Infrastructure, which it defines as:

Critical Infrastructure. *Systems and assets, whether physical or virtual, so vital to the United States that the incapacity or destruction of such systems and assets would have a debilitating impact on security, national economic security, national public health or safety, or any combination of those matters. (Source: §1016(e) of the USA Patriot Act of 2001 (42 U.S.C. §5195c(e))*

Within the context of NIPP 2013, security is interwoven with resilience and another Key Concept presented is that both security and resilience are strengthened through risk management, which will be discussed later in the review.

FEMA's *National Disaster Recovery Framework* (NDRF) uses a similar definition of resilience:

Resilience—*Ability to adapt to changing conditions and withstand and rapidly recover from disruption due to emergencies.*

Within the context of the NDRF, sustainability is partnered with resilience. One of the Core Principles identified in the NDRF is titled *Resilience and Sustainability:*

A successful recovery process promotes practices that minimize the community's risk to all hazards and strengthens its ability to withstand and recover from future disasters, which constitutes a community's resiliency. A successful recovery process engages in a rigorous assessment and understanding of risks and vulnerabilities that might endanger the community or pose additional recovery challenges. The process promotes implementation of the National Infrastructure Protection Plan (NIPP) risk management framework to enhance the resilience and protection of critical infrastructure against the effects of future disasters. Resilience incorporates hazard mitigation and land use planning strategies; critical infrastructure, environmental and cultural resource protection; and sustainability practices to reconstruct the built environment, and revitalize the economic, social and natural environments.

Understanding how these definitions could apply to NFPA's mission is clarified in planning efforts such as the work of The Infrastructure Security Partnership (TISP). In *Regional Disaster Resilience: A Guide to Developing an Action Plan* 2011 Edition, TISP couples resilience specifically with disasters and infrastructure:

Disaster Resilience (for regions and communities): Capability to prepare for, prevent, protect against, respond or mitigate any anticipated or unexpected significant threat or event, including terrorist attacks, and to adapt to changing conditions and rapidly recover and reconstitute critical assets, operations, and services with minimum damage and disruption to public health and safety, the economy, environment, and national security.

Infrastructure Resilience: The ability to resist, absorb, and recover from or successfully adapt to adversity or a change in conditions; capacity to recognize threats and hazards and make adjustments that will improve future protection efforts and risk reduction measures.

The National Infrastructure Advisory Council offers its definition of infrastructure resilience in its September 8, 2009 report:

Infrastructure resilience is the ability to reduce the magnitude and/or duration of disruptive events. The effectiveness of a resilient infrastructure or enterprise depends upon its ability to anticipate, absorb, adapt to, and/or rapidly recover from a potentially disruptive event.

Similarly, the Disaster Roundtable of the National Research Council in the report of its 12th meeting, *Creating a Disaster Resilient America: Grand Challenges in Science and Technology,* refers to disaster resilience communities:

Disaster resilience communities can be understood as those that have the capacity to take the requisite mitigation and preparedness actions to withstand extreme natural or human events. Resiliency embodies four basic dimensions of society-the technical, organizational, social and the economic.

The common themes from these many definitions can be summarized for their use in characterizing the role of NFPA codes and standards as follows:

1. Resilience includes *technical, organizational, social and economic* dimensions.
2. Resilience requires actions described as planning, preparing, preventing, protecting, mitigating and responding.
3. Resilience requires preparation and response to be adaptive.
4. Resilience should focus on minimizing *damage and disruption to public health and safety, the economy, environment, and national security.*
    Specific to design standards activities, a fifth theme can be added regarding continuity of functionality:
5. Resilience includes the ability of structures and systems to withstand these external events, whether natural or human-created.

Performance Goals and Objectives

NIPP 2013 provides a clear and useful set of goals regarding the strengthening of the security and resilience of the national critical infrastructure. However, these goals apply equally well to the broader applications of disaster resilience. Likewise, these goals speak to NFPA's mission.

Assess and analyze threats to, vulnerabilities of, and consequences to critical infrastructure to inform risk management activities;

Secure critical infrastructure against human, physical, and cyber threats through sustainable efforts to reduce risk, while accounting for the costs and benefits of security investments;

Enhance critical infrastructure resilience by minimizing the adverse consequences of incidents through advance planning and mitigation efforts, and employing effective responses to save lives and ensure the rapid recovery of essential services;

Share actionable and relevant information across the critical infrastructure community to build awareness and enable risk-informed decision making; and

Promote learning and adaptation during and after exercises and incidents.

In conjunction with these goals, NIPP-2013 identifies seven Core Tenets, three of which deal with identifying, understanding, and managing risk. The seventh deals with design consideration, which could be reflected in codes and standards:

Security and resilience should be considered during the design of assets, systems, and networks. *As critical infrastructure is built and refreshed, those involved in making design decisions, including those related to control systems, should consider the most effective and efficient ways to identify, deter, detect, disrupt, and prepare for threats and hazards; mitigate vulnerabilities; and minimize consequences. This includes considering infrastructure resilience principles.*

As will be evident from the discussion of Frameworks for Resilience below, the establishment of specific performance goals and objectives will be a critical first step in the process of pursuing Community or Disaster Resilience. NFPA has both documents and processes that could be applied for defining these goals and objectives. *Community Resilience Planning Guide for Buildings and Infrastructure Systems,* (CRPG), NIST Special Publication 1190, catalogs performance objectives specific to buildings, based on how they need to function during and after an event. These categories will help inform design criteria and also emergency planning.

*Category A—**Safe and operational**: Essential facilities such as hospitals and emergency operations centers*

*Category B—**Safe and usable during repair**: "shelter in place" residential buildings, neighborhood businesses and services and buildings needed for emergency operations*

*Category C—**Safe and not usable**: The minimum needed to save lives. These Facilities may be repaired or replaced as needed to restore the economy*

*Category D—**Unsafe- partial or complete collapse**: damage that will lead to casualties*

Frameworks for Resilience

As referenced above with FEMA's NDRF, the concepts of *frameworks* for developing or enhancing resilience are common in the literature. Because of the complexity and interconnection of the many elements identified in the definitions outlined above, *frameworks* are presented to provide a means of facilitating successful interactions. Four references are discussed here. The first is NIPP-2013. The framework provided in Sect. 5, *Collaborating to Manage Risk,* of this reference focuses on a general approach to making risk-informed decisions.

This section is organized based on the critical infrastructure risk management framework, introduced in the 2006 NIPP and updated in this National Plan. The updates help to clarify the components and streamline the steps of the framework, depicted in Fig. 3 below.

Specifically, the three elements of critical infrastructure (physical, cyber, and human) are explicitly identified and should be integrated throughout the steps of the framework, as appropriate. In addition, the updated framework consolidates the number of steps or "chevrons" by including prioritization with the implementation of risk management activities. Prioritization of risk mitigation options is an integral part of the decision-making process to select the risk management activities to be implemented. Finally, a reference to the feedback loop is removed and instead, the framework now depicts the importance of information sharing throughout the entire risk management process. Information is shared through each step of the framework, to include the "measure effectiveness" step, facilitating feedback and enabling continuous improvement of critical infrastructure security and resilience efforts.

The need for flexibility and the connection to the *Comprehensive Preparedness Guide* (CPG) 201 is also addressed in NIPP 2013:

The critical infrastructure risk management framework is designed to provide flexibility for use in all sectors, across different geographic regions, and by various partners. It can be tailored to dissimilar operating environments and applies to all threats and hazards. The risk management framework is intended to complement and support completion of the Threat and Hazard Identification and Risk Assessment (THIRA) process as conducted by regional, SLTT, and urban area jurisdictions to establish capability priorities. Comprehensive Preparedness Guide 201: Threat and Hazard Identification and Risk Assessment, Second Edition cites infrastructure owners and operators as sources of threat and hazard information and as valuable partners when completing the THIRA process.

That second reference, *CPG 201: Threat and Hazard Identification and Risk Assessment*, Second Edition, as referenced in NIPP 2013, describes a four-step process for developing a risk assessment. These four steps are as follows:

1. **Identify the Threats and Hazards of Concern**. *Based on a combination of experience, forecasting, subject matter expertise, and other available resources, identify a list of the threats and hazards of primary concern to the community.*
2. **Give the Threats and Hazards Context**. *Describe the threats and hazards of concern, showing how they may affect the community.*
3. **Establish Capability Targets**. *Assess each threat and hazard in context to develop a specific capability target for each core capability identified in the National Preparedness Goal. The capability target defines success for the capability.*
4. **Apply the Results**. *For each core capability, estimate the resources required to achieve the capability targets through the use of community assets and mutual aid, while also considering preparedness activities, including mitigation opportunities.*

CPG 201 also refers to five preparedness missions that reflect the concepts of resilience identified above:

Prevention: *Prevent, avoid, or stop an imminent, threatened, or actual act of terrorism.*

Protection: *Protect our citizens, residents, visitors, and assets against the greatest threats and hazards in a manner that allows our interests, aspirations, and way of life to thrive.*

Mitigation: Reduce the loss of life and property by lessening the impact of future disasters.

Response: Respond quickly to save lives; protect property and the environment; and meet basic human needs in the aftermath of a catastrophic incident.

Recovery: Recover through a focus on the timely restoration, strengthening, and revitalization of infrastructure, housing, and a sustainable economy, as well as the health, social, cultural, historic, and environmental fabric of communities affected by a catastrophic incident.

Although Community Resilience is only listed as a core capability under Mitigation in CPG 201, all of the preparedness missions are integral to both Community Resilience and Disaster Resilience as defined above.

The third framework appears in the *NDRF*. As the name indicates, this framework focuses on the Recovery aspects of resilience. The document is intended for broad application across diverse stakeholders. Of particular relevance to NFPA codes and standards is the emphasis on pre-planning:

Recovery begins with pre-disaster preparedness and includes a wide range of planning activities. The NDRF clarifies the roles and responsibilities for stakeholders in recovery, both pre- and post-disaster. It recognizes that recovery is a continuum and that there is opportunity within recovery. It also recognizes that when a disaster occurs, it impacts some segments of the population more than others.

The ability of a community to accelerate the recovery process begins with its efforts in pre-disaster preparedness, mitigation and recovery capacity building. These efforts result in a resilient community with an improved ability to withstand, respond to and recover from disasters. Timely decisions in response to disaster impacts can significantly reduce recovery time and cost.

The NDRF describes key principles and steps for community recovery planning and implementation. It promotes a process in which the impacted community fully engages and considers the needs of all its members. A key element of the process is that the impacted community assumes the leadership in developing recovery priorities and activities that are realistic, well-planned and clearly communicated.

This pre-planning is reiterated under Core Principles in the NDRF:

The speed and success of recovery can be greatly enhanced by establishment of the process and protocols prior to a disaster for coordinated post-disaster recovery planning and implementation. All stakeholders should be involved to ensure a coordinated and comprehensive planning process, and develop relationships that increase post-disaster collaboration and unified decision making. Another important objective of pre-disaster recovery planning is to take actions that will significantly reduce disaster impacts through disaster-resilient building practices....

NDRF includes several informative concepts for the codes and standards process. First, in its description of the Recovery Continuum, NDRF provides an excellent guide for establishing priorities that could be addressed in NFPA documents. This figure is contained in Appendix 2. The figure titled Recovery Continuum—Description of Activities by Phase identifies four phases of recovery, Preparedness (Ongoing), Short-Term (days), Intermediate (Weeks-Months), and Long-Term (Months-Years). The importance of health care, sheltering, and mitigation activities as immediate Short-

Term priorities should influence the focus of NFPA activities. Likewise, the Long-Term emphasis on *Resilient Rebuilding* expands on the *disaster-resilient building practices* referenced above and identifies two items important to NFPA's mission:

> *The recovery is an opportunity for communities to rebuild in a manner which reduces or eliminates risk from future disasters and avoids unintended negative environmental consequences.*
>
> *Communities incorporate stronger building codes and land use ordinances. Vulnerable structures are retrofitted, elevated or removed from harm.*

The fourth framework is being developed by NIST. The President's Climate Action Plan of June 2013 included the NIST initiative described below:

> **Boosting the Resilience of Buildings and Infrastructure:** *The National Institute of Standards and Technology will convene a panel on disaster-resilience standards to develop a comprehensive, community-based resilience framework and provide guidelines for consistently safe buildings and infrastructure—products that can inform the development of private-sector standards and codes.*

Using a series of workshops with a cross section of stakeholders, NIST developed a comprehensive *Disaster Resilience* Framework for achieving community resilience that considers the interdependence of the community's physical and human assets, operations, and policies/regulations. The expectation for this document is that it will include the following:

- *Define community-based disaster resilience for the built environment*
- *Identify consistent performance goals and metrics for buildings and infrastructure and lifeline systems to enhance community resilience,*
- *Identify existing standards, codes, guidelines, and tools that can be implemented to enhance resilience, and*
- *Identify gaps in current standards, codes, and tools that if successfully addressed, can lead to enhanced resilience.*

The emphasis of this effort parallels and supports those issues identified in NIPP 2013 for the *built environment*. The draft also introduces a sample resilience matrix that characterizes the hazard levels as *routine, expected, and extreme*. It also characterizes the time dependence of the response and recovery in three phases similar to the post event phases identified in the NDRF Disaster Recovery Continuum. This proposed framework also introduces a set of performance goals, *Categories*, as described above.

Appendix 2
Recovery Continuum

Recovery Continuum – Description of Activities By Phase

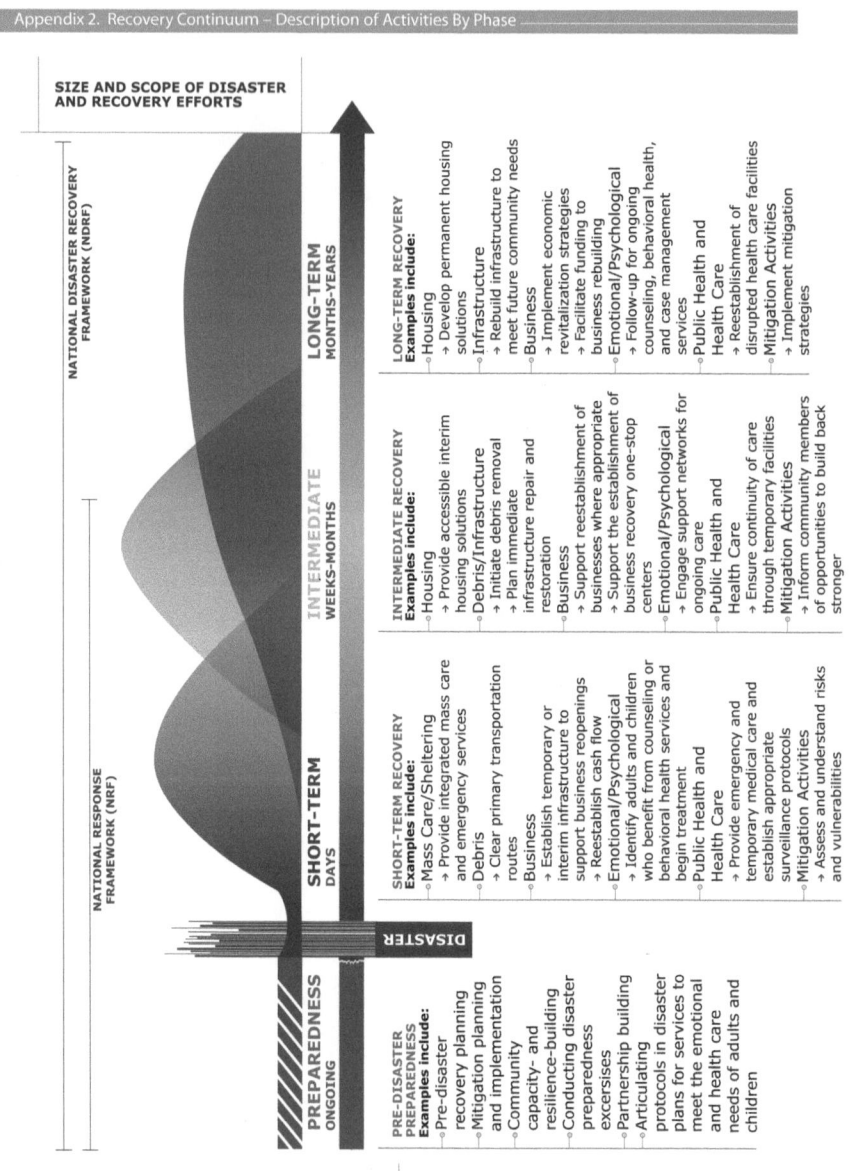

Appendix 2. Recovery Continuum – Description of Activities By Phase

SIZE AND SCOPE OF DISASTER AND RECOVERY EFFORTS

NATIONAL DISASTER RECOVERY FRAMEWORK (NDRF)

NATIONAL RESPONSE FRAMEWORK (NRF)

DISASTER

PREPAREDNESS — ONGOING

PRE-DISASTER PREPAREDNESS Examples include:
- Pre-disaster recovery planning
- Mitigation planning and implementation
- Community capacity- and resilience-building
- Conducting disaster preparedness exercises
- Partnership building
- Articulating protocols in disaster plans for services to meet the emotional and health care needs of adults and children

SHORT-TERM — DAYS

SHORT-TERM RECOVERY Examples include:
- Mass Care/Sheltering
  → Provide integrated mass care and emergency services
- Debris
  → Clear primary transportation routes
- Business
  → Establish temporary or interim infrastructure to support business reopenings
  → Reestablish cash flow
- Emotional/Psychological
  → Identify adults and children who benefit from counseling or behavioral health services and begin treatment
- Public Health and Health Care
  → Provide emergency and temporary medical care and establish appropriate surveillance protocols
- Mitigation Activities
  → Assess and understand risks and vulnerabilities

INTERMEDIATE — WEEKS-MONTHS

INTERMEDIATE RECOVERY Examples include:
- Housing
  → Provide accessible interim housing solutions
- Debris/Infrastructure
  → Initiate debris removal
  → Plan immediate infrastructure repair and restoration
- Business
  → Support reestablishment of businesses where appropriate
  → Support the establishment of business recovery one-stop centers
- Emotional/Psychological
  → Engage support networks for ongoing care
- Public Health and Health Care
  → Ensure continuity of care through temporary facilities
- Mitigation Activities
  → Inform community members of opportunities to build back stronger

LONG-TERM — MONTHS-YEARS

LONG-TERM RECOVERY Examples include:
- Housing
  → Develop permanent housing solutions
- Infrastructure
  → Rebuild infrastructure to meet future community needs
- Business
  → Implement economic revitalization strategies
  → Facilitate funding to business rebuilding
- Emotional/Psychological
  → Follow-up for ongoing counseling, behavioral health, and case management services
- Public Health and Health Care
  → Reestablishment of disrupted health care facilities
- Mitigation Activities
  → Implement mitigation strategies

This recovery continuum describes overlapping recovery activities by phase.